POSTMO

Einstein and the
Birth of Big Science

Peter Coles

Series editor: Richard Appignanesi

ICON BOOKS UK

TOTEM BOOKS USA

Published in the UK in 2000
by Icon Books Ltd., Grange Road,
Duxford, Cambridge CB2 4QF
email: info@iconbooks.co.uk
www.iconbooks.co.uk

Published in the USA in 2001
by Totem Books
Inquiries to: Icon Books Ltd.,
Grange Road, Duxford,
Cambridge CB2 4QF, UK

Distributed in the UK, Europe,
Canada, South Africa and Asia
by the Penguin Group:
Penguin Books Ltd.,
27 Wrights Lane,
London W8 5TZ

In the United States,
distributed to the trade by
National Book Network Inc.,
4720 Boston Way, Lanham,
Maryland 20706

Library of Congress catalog
card number applied for

Published in Australia in 2001
by Allen & Unwin Pty. Ltd.,
PO Box 8500, 9 Atchison Street,
St. Leonards, NSW 2065

Text copyright © 2000 Peter Coles

The author has asserted his moral rights.

Series editor: Richard Appignanesi

ISBN 1 84046 183 7

Typesetting by Wayzgoose

Printed and bound in the UK by
Cox & Wyman Ltd., Reading

The Age of Big Science

Science has developed at an astonishing rate over the past hundred years. All of the major scientific disciplines – physics, chemistry and biology – are now unrecognisable compared to what they were in 1900. Applications of the new scientific ideas, for example in technology and medicine, have altered everyday life to an equally remarkable extent, at least in the developed world. In many cases these developments have been for the common good, although they have sometimes resulted in great social change and have left large sections of society behind.

Science and technology have also affected the way many people think about the world. For some, science has dispensed with the need for religion. For others it *is* a religion, one whose litany is rich with mystifying jargon. Moreover, as science has developed, scientists have begun to think about aspects of nature that have traditionally been covered by religious teaching. This is just one example of science moving into and disturbing what had previously been a separate arena of human thought.

In the physical sciences, developments have been particularly remarkable. Physicists have unravelled the structure of matter on the tiniest accessible scales, breaking up atomic nuclei into elementary particles and studying the forces that cause these particles to interact. Astronomers discovered in this century that the Universe is expanding, and cosmologists are now trying to understand the very instant of creation at the Big Bang that started this expansion off. These daring adventures of the mind are based on foundations of experiment, observation and theory. But the more ambitious scientists become, the further their theoretical ideas lie beyond the grasp of the general public. Likewise, their experiments and observations become more and more expensive to make, and more and more difficult to subject to independent test. 'Big Science' has become the preserve of a very few specialists, distancing it even further from popular understanding than science generally.

The modern era of scientific thought began with Galileo and Newton. Since then, science has been getting bigger and bigger. But the early years of the 20th century saw the birth of Big Science proper. In 1919, an experiment was performed that was intended to test Einstein's general theory of relativity (see 'Key Ideas' at the end of this book). The results caused a media sensation and made Albert Einstein into a household name. As we shall see, the story of this experiment and its aftermath is an interesting case study that reveals insights into the relationship between science – specifically Big Science – and wider society. But let's start with the background to the experiment and its scientific importance.

Universal Gravitation

To a physicist, gravity is one of the fundamental forces of nature. It represents the universal tendency of all matter to attract all other matter. This universality sets it apart from, for example, the forces between electrically-charged bodies, because electrical charges can be of two different kinds, positive or negative. While electrical forces can lead either to attraction (between unlike charges) or repulsion (between like charges), gravity is always attractive.

In many ways, the force of gravity is extremely

weak. Most material bodies are held together by electrical forces between atoms which are many orders of magnitude stronger than the gravitational forces between them. But, despite its weakness, gravity is the driving force in astronomical situations because astronomical bodies, with very few exceptions, always contain exactly the same amount of positive and negative charge, and therefore never exert forces of an electrical nature on each other.

One of the first great achievements of theoretical physics was Isaac Newton's theory of universal gravitation, which unified what, at the time, seemed to be many disparate physical phenomena. Newton's theory of mechanics is encoded in three simple laws:

1. Every body continues in a state of rest or uniform motion in a straight line unless it is compelled to change that state by forces impressed upon it.
2. Rate of change of momentum is proportional to the impressed force, and is in the direction in which this force acts.

3. To every action, there is always opposed an equal reaction.

These three laws of motion are general, applying just as accurately to the behaviour of balls on a billiard table as to the motion of the heavenly bodies. All that Newton needed to do was to figure out how to describe the force of gravity. Newton realised that a body orbiting in a circle, like the Moon going around the Earth, is experiencing a force in the direction of the centre of motion (as does a weight tied to the end of a piece of string when it is twirled around one's head). Gravity could cause this motion, in the same way as it could cause apples to fall to Earth from trees. In both of these situations, the force has to be towards the centre of the Earth. Newton realised that the right form of mathematical equation was an 'inverse-square' law: 'the attractive force between any two bodies depends on the product of the masses of the bodies and upon the square of the distance between them.'

It was a triumph of Newton's theory, based on the inverse-square law of universal gravitation, that it could explain the laws of planetary motion obtained by Johannes Kepler more than a century earlier. So spectacular was this success that the idea of a Universe guided by Newton's laws of motion was to dominate scientific thinking for more than two centuries. Until, in fact, the arrival on the scene of an obscure patent clerk by the name of Albert Einstein.

The Einstein Revolution

Albert Einstein was born in Ulm (Germany) on 14 March 1879, but his family soon moved to Munich, where he spent his school years. The young Einstein was not a particularly good student, and in 1894 he dropped out of school entirely when his family moved to Italy. After failing the entrance examination once, he was eventually admitted to the Swiss Institute of Technology in Zurich in 1896. Although he did fairly well as a student in Zurich, he was unable to get a job in any Swiss university, as he was held to be extremely lazy. He left academia to work in the Patent Office at Bern in 1902. This gave him a good wage and, since the tasks given to a junior patent clerk were not exactly onerous, it also gave him plenty of spare time to think about physics.

Einstein's special theory of relativity was published in 1905. It stands as one of the greatest intellectual achievements in the history of human thought. It is made even more remarkable by the fact that Einstein was still working as a patent clerk at the time, and was only doing physics as a kind of hobby. What's more, he also published seminal works that year on the photoelectric effect (which was to inspire many developments in quantum theory) and on the phenomenon of Brownian motion (the jiggling of microscopic particles as they are buffeted by atomic collisions). But the reason why the special theory of relativity stands head-and-shoulders above his own work of this time, and that of his colleagues in the world of mainstream physics, is that Einstein managed to break away completely from the concept of time as an absolute property that marches on at the same rate for everyone and everything. This idea is built into the Newtonian picture of the world, and most of us regard it as being so obviously true that it does not bear discussion. It takes a genius to break down conceptual barriers of such magnitude.

The idea of relativity did not originate with Einstein. The principle of it had been articulated by Galileo nearly three centuries earlier. Galileo claimed that only relative motion matters, so there could be no such thing as absolute motion. He argued that if you were travelling in a boat at constant speed on a smooth lake, then there would be no experiment that you could do in a sealed cabin on the boat that would indicate to you that you were moving at all. Of course, not much was known about physics in Galileo's time, so the kinds of experiment he could envisage were rather limited.

Einstein's version of the principle of relativity simply turned it into the statement that all laws of nature have to be exactly the same for all observers in relative motion. In particular, Einstein decided that this principle must apply to the theory of electromagnetism, constructed by James Clerk Maxwell, which describes amongst other things the forces between charged bodies mentioned above. One of the consequences of Maxwell's theory is that the speed of light (in vacuum) appears as a universal constant (usually given the symbol 'c'). Taking the principle of relativity seriously means that all observers have to measure the same value of c, whatever their state of motion. This seems straightforward enough, but the consequences are nothing short of revolutionary.

Thought Experiment (1)

Einstein decided to ask himself specific questions about what would be observed in particular kinds of experiments involving the exchange of light signals. He worked a great deal with *gedanken* (thought) experiments of this kind. For example, imagine there is a flash bulb in the centre of a railway carriage moving along a track. At each end of the carriage there is a clock, so that when the flash illuminates it we can see the time. If the flash goes off, then the light signal reaches both ends of the carriage simultaneously, from the point of view of passengers sitting in the carriage. The same time is seen on each clock.

Now picture what happens from the point of view of an observer at rest who is watching the train from the track. The light flash travels with the same speed in our reference frame as it did for the passengers. But the passengers at the back of the carriage are moving into the signal, while those at the front are moving away from it. This observer therefore sees the clock at the back of the train light up before the clock at the front does. But when the clock at the front does light up, it reads the same time as the clock at the back did! This observer has to conclude that something is wrong with the clocks on the train.

This example demonstrates that the concept of simultaneity is relative. The arrivals of the two light flashes are simultaneous in the frame of the carriage, but occur at different times in the frame of the track. Other examples of strange relativistic phenomena include time dilation (moving clocks appear to run slow) and length contraction (moving rulers appear shorter). These are all consequences of the assumption that the speed of light must be the same as measured by all observers. Of course, the examples given above are a little unrealistic. In order to show noticeable effects, the velocities concerned must be a sizeable fraction of c. Such speeds are unlikely to be reached in railway carriages. Nevertheless, experiments have been done that show that time dilation effects are real. The decay rate of radioactive particles is much slower when they are moving at high velocities because their internal clock runs slowly.

Special relativity also spawned the most famous equation in all of physics: $E=mc^2$, expressing the equivalence between matter and energy. This has also been tested experimentally, rather too often, because it is the principle behind the explosion of atomic bombs.

Remarkable though the special theory undoubtedly is, it is seriously incomplete because it deals only with bodies moving with constant velocity with respect to each other. Even Chapter 1 of the laws of nature, written by Newton, had been built around the causes and consequences of velocities that change with time. Newton's second law is about the rate of change of momentum of an object, which in layman's terms is its acceleration. Special relativity is restricted to so-called *inertial* motions, i.e. the motions of particles that are not acted upon by any external forces. This means that special relativity cannot describe accelerated motion of any kind and, in particular, cannot describe motion under the influence of gravity.

The Principle of Equivalence

Einstein had a number of deep insights in how to incorporate gravitation into relativity theory. For a start, consider Newton's theory of gravity. In this theory, the force on a particle of mass m due to another particle of mass M depends on the product of these masses and the square of the distance between the particles. According to Newton's laws of motion, this induces an acceleration in the first particle given by $F=ma$. The m in this equation is called the *inertial* mass of the particle, and it determines the particle's resistance to being accelerated. In the inverse-square law of gravity, however, the mass m measures the reaction of one particle to the gravitational force produced by the other particle. It is therefore called the *passive* gravitational mass. But Newton's third law of motion also states that if body A exerts a force on body B then body B exerts a force on body A which is equal and opposite.

This means that m must also be the *active* gravitational mass (if you like, the gravitational charge) produced by the particle. In Newton's theory, all three of these masses – the inertial mass, the active and passive gravitational masses – are equivalent. But there seems to be no reason, on the face of it, why this should be the case. Couldn't they be different?

Einstein decided that this equivalence must be the consequence of a deeper principle called *the principle of equivalence*. In his own words, this means that 'all local, freely-falling laboratories are equivalent for the performance of all physical experiments'. What this means is essentially that one can do away with gravity as a separate force of nature and regard it instead as a consequence of moving between accelerated frames of reference.

Thought Experiment (2)

To see how this is possible, imagine a lift equipped with a physics laboratory. If the lift is at rest on the ground floor, experiments will reveal the presence of gravity to the occupants. For example, if we attach a weight on a spring to the ceiling of the lift, the weight will extend the spring downwards. Next, imagine that we take the lift to the top of a building and let it fall freely. Inside the freely-falling lift there is no perceptible gravity. The spring does not extend, as the weight is always falling at the same rate as the rest of the lift, even though the lift's speed might be changing. This is what would happen if we took the lift out into space, far away from the gravitational field of the Earth. The absence of gravity therefore looks very much like the state of free-fall in response to a gravitational force. Moreover, imagine that our lift was actually in space (and out of gravity's reach), but there was a rocket attached to it. Firing the rocket would make the lift accelerate. There is no up or down in free space, but let us assume that the rocket is attached so that the lift would accelerate in the opposite direction from before, i.e. in the direction of the ceiling.

What happens to the spring? The answer is that the acceleration makes the weight move in the reverse direction relative to the lift, thus extending the spring towards the floor. (This is like what happens when a car suddenly accelerates – the

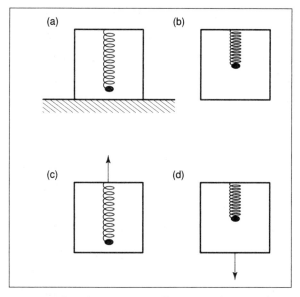

Figure 1. Thought-experiment illustrating the equivalence principle. A weight is attached to a spring, which is attached to the ceiling of a lift. In (a) the lift is stationary, but a gravitational force acts downwards; the spring is extended by the weight. In (b) the lift is in deep space, away from any sources of gravity, and is not accelerated; the spring does not extend. In (c) there is no gravitational field, but the lift is accelerated upwards by a rocket; the spring is extended. The acceleration in (c) produces the same effect as the gravitational force in (a). In (d) the lift is freely-falling in a gravitational field, accelerating downwards so no gravity is felt inside; the spring does not extend because in this case the weight is weightless and the situation is equivalent to (b).

17

passenger's head is flung backwards.) But this is just like what happened when there was a gravitational field pulling the spring down. If the lift carried on accelerating, the spring would remain extended, just as if it were not accelerating but placed in a gravitational field. Einstein's idea was that these situations do not merely appear similar: *they are completely indistinguishable.* Any experiment performed in an accelerated lift in space would give exactly the same results as one performed in a lift upon which gravity is acting. To complete the picture, now consider a lift placed inside a region where gravity is acting, but which is allowed to fall freely in the gravitational field. Everything inside becomes weightless, and the spring is not extended. This is equivalent to the situation in which the lift is at rest and where no gravitational forces are acting. A freely-falling observer has every reason to consider himself to be in a state of inertial motion.

The General Theory of Relativity

Einstein now knew how he should construct the general theory of relativity. But it would take him another ten years to produce the theory in its final form. What he had to find was a set of laws that could deal with any form of accelerated motion and any form of gravitational effect. To do this he had to learn about sophisticated mathematical techniques, such as tensor analysis and Riemannian geometry, and to invent a formalism that was truly general enough to describe all possible states of motion. He got there, but clearly it wasn't easy. While his classic papers of 1905 were characterised by brilliant clarity of thought and economy of mathematical calculation, his later work is mired in technical difficulty. People have argued that Einstein grew up as a scientist while he was developing the general theory. If so, it was obviously a difficult process for him.

Understanding the technicalities of the general theory of relativity is a truly daunting task, and calculating anything useful using the full theory is beyond all but the most dedicated specialists. While the application of Newton's theory of gravity requires one equation to be solved, Einstein's theory has no less than ten, which must all be solved simultaneously. And each separate equation is much more complicated than Newton's simple inverse-square law. Because of the equivalence between mass and energy embodied in special relativity through $E=mc^2$, all forms of energy gravitate. The gravitational field produced by a body is itself a form of energy, and it also therefore gravitates. This kind of chicken-and-egg problem is called 'non-linearity' by physicists, and it often leads to unmanageable mathematical complexity when it comes to solving the equations.

Even on a conceptual level, the theory is difficult to grasp. The relativity of time embodied in the special theory is present in the general theory, but there are additional effects of time dilation and length contraction due to gravitational effects. And the problems don't end with time! In the special theory, space at least is well-behaved. In the general theory, even this goes out of the window. Space is *curved*.

The Bending of Light

The idea that space could be warped is so difficult to grasp that even physicists don't really try to visualise such a thing. Our understanding of the geometrical properties of the natural world is based on the achievements of generations of Greek mathematicians, notably the formalised system of Euclid – Pythagoras' theorem, parallel lines never meeting, the sum of the angles of a triangle adding up to 180 degrees, and so on. All of these rules find their place in the canon of Euclidean geometry. But these laws and theorems are not just abstract mathematics. We know from everyday experience that they describe the properties of the physical world extremely well. Euclid's laws are used every day by architects, surveyors, designers and cartographers – anyone, in fact, who is concerned with the properties of shape, and the positioning of objects in space. Geometry is real.

It seems self-evident, therefore, that these properties of space that we have grown up with should apply beyond the confines of our buildings and the lands we survey. They should apply to the Universe as a whole. Euclid's laws must be built into the fabric of the world. Or must they? Although Euclid's laws are mathematically elegant and logically compelling, they are not the only set of rules that can be used to build a system of geometry. Mathematicians of the 19th century, such as Gauss and Riemann, realised that Euclid's laws represent only a special case of geometry wherein space is flat. Different systems can be constructed in which these laws are violated.

Consider, for example, a triangle drawn on a flat sheet of paper. Euclid's theorems apply here, so the sum of the internal angles of this triangle must be 180 degrees (equivalent to two right-angles). But now think about what happens if you draw a triangle on a sphere instead. It is quite possible to draw a triangle on a sphere that has *three* right angles in it. For example, draw one point at the 'north pole' and two on the 'equator' separated by one quarter of the circumference. These three points form a triangle with three right angles that violates Euclidean geometry.

Thinking this way works fine for two-dimensional geometry, but our world has three dimensions of space. Imagining a three-dimensional curved surface is much more difficult. But in any case it is probably a mistake to think of 'space' at all. After all, one can't measure space. What one can measure are distances between objects located in space using rulers or, more realistically in an astronomical context, light beams. Thinking of space as a flat or curved piece of paper encourages one to think of it as a tangible thing in itself, rather than simply as where the tangible things are not. Einstein always tried to avoid dealing with entities such as 'space' whose category of existence was unclear. He preferred to reason instead about what an observer could actually measure with a given experiment.

Following this lead, we can ask what kind of path light rays follow according to the general theory of relativity. In Euclidean geometry, light travels on straight lines. We can take the straightness of light paths to mean essentially the same thing as the flatness of space. In special relativity, light also travels on straight lines, so space is flat in this view of the world too. But remember that the general theory applies to accelerated motion, or motion in the presence of gravitational effects. What happens to light in this case?

Let us go back to the thought experiment

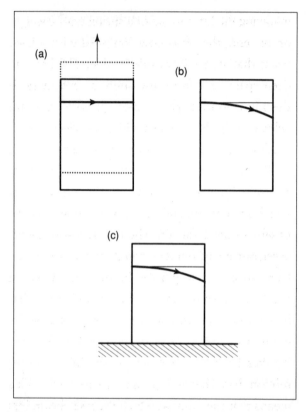

Figure 2. The bending of light. In (a), our lift is accelerating upwards, as in Figure 1(c). Viewed from outside, a laser beam follows a straight line. In (b), viewed inside the lift, the light beam appears to curve downwards. The effect in a stationary lift situated in a gravitational field is the same, as we see in (c).

involving the lift. Instead of a spring with a weight on the end, the lift is now equipped with a laser beam that shines from side to side. The lift is in deep space, far from any sources of gravity. If the lift is stationary, or moving with constant velocity, then the light beam hits the side of the lift exactly opposite to the laser device that produces it. This is the prediction of the special theory of relativity. But now imagine the lift has a rocket which switches on and accelerates it upwards. An observer outside the lift who is at rest sees the lift accelerate away, but if he could see the laser beam from outside, it would still be straight. He is not accelerating, so the special theory applies to what he sees. On the other hand, a physicist inside the lift notices something strange. In the short time it takes light to travel across, the lift's state of motion has changed (it has accelerated). This means that the point at which the laser beam hits the other wall is slightly below the starting point on the other side. What has happened is that the acceleration has 'bent' the light ray downwards.

Now remember the case of the spring and the equivalence principle. What happens when there is no acceleration but there is a gravitational field, is exactly the same as in an accelerated lift. Consider now a lift standing on the Earth's surface. The light ray must do exactly the same thing as in the accelerating lift: it bends downward. The conclusion we are led to is that gravity bends light. And if light paths are not straight but bent, then space is not flat but curved.

Newton and Soldner

The story so far gives the impression that nobody before Einstein considered the possibility that light could be bent. In fact, this is not the case. It had been reasoned before, by none other than Isaac Newton himself, that light might be bent by a massive gravitating object. In a rhetorical question posed in his *Opticks*, Newton wrote:

Do not Bodies act upon Light at a distance, and by their action bend its Rays; and is not this action . . . strongest at the least distance?

In other words, he was arguing that light rays themselves should feel the force of gravity according to the inverse-square law. As far as we know, however, he never attempted to apply this idea to anything that might be observed. Newton's query was addressed in 1801 by Johann Georg von Soldner. His work was motivated by the desire to know whether the bending of light rays might require certain astronomical observations to be adjusted. He tackled the problem using Newton's corpuscular theory of light, in which light rays consist of a stream of tiny particles. It is clear that if light does behave in this way, then the mass of each particle must be very small. Soldner was able to use Newton's theory

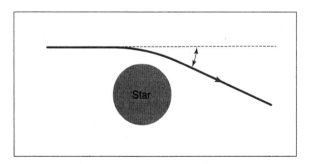

Figure 3. The ballistic scattering problem. A small body passing close to a massive one is deflected through a small angle on its way.

of gravity to solve an example of what modern-day physicists call a 'ballistic scattering problem'.

A small particle moving past a large gravitating object feels a force from the object that is directed towards the centre of the large object. If the particle is moving fast, so that the encounter does not last very long, and the mass of the particle is much less than the mass of the scattering body, what happens is that the particle merely receives a sideways kick which slightly alters the direction of its motion. The size of the kick, and the consequent scattering angle, is quite easy to calculate because the situation allows one to ignore the motion of the scatterer. Although the two bodies exert equal and opposite forces on each other, according to Newton's third law, the fact that the scatterer has a much larger mass than the 'scatteree' means that the former's acceleration is very much lower. This kind of scattering effect is exploited by interplanetary probes, which can change course without firing booster rockets by using the gravitational 'slingshot' supplied by the Sun or larger planets.

Unfortunately, this calculation has a number of

problems associated with it. Chief amongst them is the small matter that light does not actually possess mass at all. Although Newton had hit the target with the idea that light consists of a stream of particles, these photons, as they are now called, are known to be massless. Newton's theory simply cannot be applied to massless particles: they feel no gravitational force (because the force depends on their mass) and they have no inertia. What photons do in a Newtonian world is really anyone's guess. Nevertheless, the Soldner result is usually called the Newtonian prediction, for want of a better name.

Unaware of Soldner's calculation, in 1907 Einstein began to think about the possible bending of light. By this stage, he had already arrived at the equivalence principle, but it was to be another eight years before the general theory of relativity was completed. He realised that the equivalence principle in itself required light to be bent by gravitating bodies. But he assumed that the effect was too small ever to be observed in practice, so he shelved the calculation. In 1911, still before the general theory was ready, he

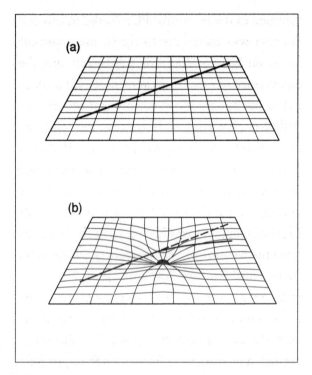

Figure 4. Curved space and the bending of light. In this illustration, space is represented as a two-dimensional surface. In the absence of any gravitating bodies, light travels in a straight line like a ball rolling on a smooth, flat table-top (a). When a massive body is placed in the way, space becomes curved: the closer you get to the body, the more curved it is (b). The effect on light is as if the ball were rolling across a table-top with a dip in the middle: it is deflected away from a straight line.

returned to the problem. What he did in this calculation was essentially to repeat the argument based on Newtonian theory, but exploiting the equivalence of matter (m) and energy (E) embodied in his most famous equation of all, $E=mc^2$. Although photons don't have mass, they certainly have energy, and Einstein's theory says that even pure energy has to behave in some ways like mass. Using this argument, and spurred on by the realisation that the light deflection he was thinking about might after all be measurable, he calculated the bending of light from background stars by the Sun. For light just grazing the Sun's surface, the answer was 0.87 seconds of arc. (One arc second is 1/3600th of one degree; for reference, the angle in the sky occupied by the Sun is around half a degree.) This answer is precisely the same as the Newtonian value obtained more than a century earlier by Soldner.

The predicted deflection is tiny, but according to the astronomers Einstein consulted, it could just about be measured. Stars appearing close to the Sun would appear to be in slightly different positions in the sky than they would be when the

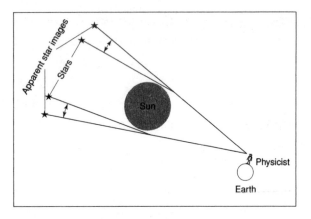

Figure 5. The bending of light by the Sun. Light from background stars follows paths like that shown in Figure 3. The result is that the stars are seen in slightly different positions in the sky when the Sun is in front of them, compared to their positions when the Sun is elsewhere.

Sun was in another part of the sky. It was hoped that this kind of observation could be used to test Einstein's theory. The only problem was that the Sun would have to be edited out of the picture, otherwise stars would not be visible close to it at all. In order to get around this problem, the measurement would have to be made at a very special time and place: during a total eclipse of the Sun.

The Anatomy of an Eclipse

Our journey is going to take us into the realm of fundamental physics, so it is well to start close to home with a description of what an eclipse actually is. As we are all taught at school, the Earth travels in a roughly circular path around the Sun, which is some 93 million miles away. It takes about 365 days to complete each circuit. The Earth also spins on an axis that passes through its centre, from the North pole to the South. It takes roughly 24 hours to complete one rotation. The Earth's rotation is why the Sun appears to rise in the morning and set in the evening.

The Moon, on the other hand, travels in a similarly circular path, not around the Sun but around the Earth. It takes about 28 days to complete one of its circuits, roughly defining the length of a month. During each month, the relative orientation of the Earth, Moon and Sun changes so that the part of the Moon's surface which is facing the Sun, and which we can also see from Earth, changes during the course of a month. We observe the Moon by the sunlight reflected from its surface – this causes the phases of the Moon. The Moon is 'full' when it reflects the Sun's light full-on, and 'new' when the part lit by the Sun is hidden from our view. In between, we see partial illumination of the lunar surface in a variety of crescent shapes, depending on the angle between us, Sun and Moon. You can demonstrate the entire repertoire of these phases at home, using a tennis ball and a flashlight.

It is interesting that the Moon's orbit around the Earth lies in roughly the same plane as the Earth's orbit around the Sun, almost as if these motions were drawn on a flat piece of paper. It therefore happens from time to time that the Moon, Sun and Earth form a straight line. This can happen either with the Earth between Moon and Sun, or with the Moon between Earth and Sun.

In the first of these cases, the Moon passes into a region of shadow cast by the Earth – a phenomenon called a *lunar eclipse*. Because the Moon must be opposite the Sun in the sky for this to happen, this only happens close to, or during, a full moon. Lunar eclipses are quite dramatic things to see: the moon takes on a strange orangey-red appearance and grows eerily dim.

But this is small potatoes compared to what happens in the second case, a *solar eclipse*.

What happens in a solar eclipse is that the Moon casts a shadow somewhere on the surface of Earth which, if you could observe it from space, would be a roughly oval splotch a few hundred miles wide. As the Earth rotates, and the Moon also moves across the line joining the Earth and Sun, the shadowy splotch moves, sweeping out a path on the Earth's surface known as the *path of totality*. Viewed from the Earth, an observer somewhere on the path of totality would see the Moon's disk pass in front of the Sun, progressively hiding more and more of it from view. Just as a lunar eclipse can only occur close to a full moon, a solar eclipse must happen close to a new moon, so the moon will generally not be particularly easy to see before the eclipse. Hence the rather unexpected arrival of an eclipse to those who are unprepared. If the observer is in the right place at the right time, the Sun's disk may be completely obscured by the Moon, in which case the eclipse is a *total eclipse*. Eventually, as the shadow moves along the path of totality, places behind it move back into daylight, while places ahead of it move into darkness. The path of totality can extend for many thousands of miles, and the duration of totality in any one place can be as long as about seven minutes. Either side of the path of totality one may see a *partial eclipse*, in which the Moon eats away at the Sun but never swallows it entirely.

All of this can be understood using simple geometry and knowledge of the regular orbital motions of Earth and Moon. But the possibility of having a total eclipse relies on one extraordinary and fascinating coincidence. The Moon is much closer to us than the Sun is (250,000 miles compared to 93 million), but is also very much smaller (its diameter is just over 2,000 miles, while the Sun is about 900,000 miles across). Somehow these numbers have conspired with the laws of trigonometry to produce a situation in which the apparent size of the Moon's disk is almost exactly the same as that of the Sun. There is no convincing explanation of how this remarkable coincidence came about, but without it total eclipses would be impossible. On the other hand, eclipses are not always total. The apparent sizes of Sun and Moon are only equal if the distances involved during the eclipse are exactly right. Sometimes an eclipse can occur in which the apparent size of the Moon is too small to blot out the Sun entirely. What happens in such a case is that the Moon's shadow creates the impression of a 'hole' in the Sun, rather like a Polo mint. This is called an *annular eclipse*.

During a total eclipse, one can see the parts of the Sun that other observations cannot reach. For example, the ghostly halo surrounding the eclipse itself is formed by the *corona* and *chromosphere*, which are made of extremely tenuous gas, much less dense than the Sun itself but very much hotter (the Sun's surface has a temperature of around 6,000 degrees, whereas the corona's temperature is measured in millions of degrees). But observations of these phenomena during eclipses are not as useful as one might think, because only a small part of the energy they produce is in the form of visible light. Ultra-violet and X-ray observations are much more important, but these forms of light do not penetrate the Earth's atmosphere at all, eclipse or no eclipse. Nowadays, specialised satellites carrying sophisticated detectors keep the Sun in a state of surveillance. And using an ingenious device called a coronagraph, they can create permanent artificial eclipses in order to view the corona continuously and efficiently.

Although total eclipses are rare in any one place, they are actually quite common in global terms. At least two solar eclipses must occur every year, although most of these are partial, and it often happens that a partial eclipse occurs when no part of the Earth witnesses a total eclipse. The maximum number of eclipses that can occur in a year is seven, at least two of which must be lunar.

We know so much about eclipses – when they occur, how long they will last, where the path of totality lies – because we know so much about the motions of the Earth and Moon around the Sun. This owes a great deal to centuries of accurate observation, but also depends on theory. Astronomers have a mathematical model of how the solar system works, governed by equations that enable them to calculate the relative positions of all planets and satellites with essentially perfect accuracy. The equations used to do this are more than 300 years old. They are the laws of motion and the law of universal gravitation discovered by Isaac Newton and published in his monumental *Principia* in 1686. But it is ironic that while Newton's laws are used to predict eclipses, including the eclipse of 1919, the observations that were made then would lead to the overthrow of the Newtonian view of the world.

A Crucial Correction

But this isn't quite where we take up the story of the famous eclipse expeditions of 1919. There is a twist in the tale. In 1915, with the full general theory of relativity in hand, Einstein returned to the light-bending problem. And he soon realised that in 1911 he had made a mistake. The correct answer was *not* the same as the Newtonian result, but *twice* as large. He had neglected to include all effects of curved space in the earlier calculation. Now there was real motivation to test the theory, which gave a larger answer, 1.74 arc seconds, compared to the 0.87 arc seconds obtained using Newtonian theory. Not only did that make it easier to measure, but it also offered the possibility of a definitive test of the theory.

In 1912, an Argentinian expedition had been sent to Brazil to observe a total eclipse. Light-bending measurements were on the agenda, but bad weather prevented them making any observations. In 1914, a German expedition, organised by Erwin Freundlich and funded by Krupp, the arms manufacturer, was sent to the Crimea to observe the eclipse due on 21 August. But when the First World War broke out, the party was warned off. Most returned home, but others were detained in Russia. No results were obtained. The war made further European expeditions impossible. One wonders how Einstein would have been treated by history if either of the 1912 or 1914 expeditions had been successful. Until 1915, his reputation was riding on the incorrect value of 0.87 arc seconds. As it turned out, the 1919 British expeditions to Sobral and Principe were to prove his later calculation to be right. And the rest, as they say, is history.

Eddington and the Expeditions

The story of the 1919 expeditions revolves around an astronomer by the name of Arthur Stanley Eddington. Eddington was born in Cumbria in 1882, but moved with his mother to Somerset in 1884 when his father died. He was brought up as a devout Quaker, a fact that plays an important role in the story of the eclipse expedition. In 1912, aged only 30, he became the Plumian Professor of Astronomy and Experimental Philosophy at the University of Cambridge, the most prestigious astronomy chair in Britain, and two years later he became director of the Cambridge observatories. Eddington had led an expedition to Brazil in 1912 to observe an eclipse, so his credentials made him an ideal candidate to measure the predicted bending of light.

Eddington was in England when Einstein presented the general theory of relativity to the Prussian Academy of Sciences in 1915. Since Britain and Germany were at war at that time, there was no direct communication of scientific results between the two countries. But Eddington was fortunate in his friendship with an astronomer called Willem De Sitter, later to become one of the founders of modern cosmology, and who was in neutral Holland at the time. De Sitter received copies of Einstein's papers, and wasted no time in passing them on to Eddington in 1916. Eddington was impressed by the beauty of Einstein's work, and immediately began to promote it. In a report to the Royal Astronomical Society in early 1917, he particularly stressed the importance of testing the theory using measurements of light bending. A few weeks later, the Astronomer Royal, Sir Frank Watson Dyson, realised that the eclipse of 29 May 1919 was especially propitious for this task. Although the path of totality ran across the Atlantic ocean from Brazil to West Africa, the position of the Sun at the time would be right in front of a prominent grouping of stars known as the Hyades. When totality occurred, the sky behind the Sun would be glittering with bright stars whose positions could be measured.

Dyson began immediately to investigate possible observing sites. It was decided to send not one, but two expeditions. One, led by Eddington, was to travel to the island of Principe off the coast of Spanish Guinea in West Africa, and the other, led by Andrew Crommelin (an astronomer at the Royal Greenwich Observatory), would travel to Sobral in northern Brazil. An application was made to the Government Grant Committee to fund the expeditions: £100 for instruments and £1,000 for travel and other costs. Preparations began, but immediately ran into problems.

Although Britain and Germany had been at war since 1914, conscription into the armed forces was not introduced in England until 1917. At the age of 34, Eddington was eligible for the draft, but as a Quaker he let it be known that he would refuse to serve. The climate of public opinion was heavily against conscientious objectors. Eddington might well have been sent with other Quaker friends to a detention camp and spent the rest of the war peeling potatoes. Dyson, and other prominent Cambridge academics, went to the Home Office to argue that it could not be in the nation's interest to have such an eminent scientist killed in the trenches of the Somme. After much political wrangling, a compromise was reached.

Eddington's draft was postponed, but only on condition that if the war ended by 29 May 1919, he must lead the expedition to Principe.

Even with this hurdle out of the way, significant problems remained. The expeditions would have to take specialised telescopes and photographic equipment. But the required instrument-makers had either been conscripted or were engaged in war work. Virtually nothing could be done until the armistice was signed in November 1918. Preparations were hectic, for the expeditions would have to set sail in February 1919 in order to arrive and set up camp in good time. Moreover, reference plates would have to be made. The experiment required two sets of photographs of the appropriate stars. One, of course, would be made during the eclipse, but the other set (the reference plates) had to be made when the Sun was nowhere near that part of the sky. In order to correct for possible systematic effects, the reference plates should ideally be taken at the same site and at the same elevation in the sky. This would mean waiting at the observation site until the stars that would be behind the Sun during the eclipse were at the same position in the sky before dawn. This was not too much of a problem at Sobral, where the eclipse occurred in the morning, but at Principe, Eddington would have to wait for several months to take his reference plates.

In the end, the expeditions set off on time in February 1919, and back home the astronomical community – particularly in Britain – chewed its collective fingernails. There were several possible outcomes. They might fail to measure anything, due to bad weather or some other mishap. They might measure no deflection at all, which would contradict all the theoretical ideas of the time. They might find the Newtonian value, which would humiliate Einstein. Or they might vindicate him, by measuring the crucial factor of two. Which would it be?

The June 1919 issue of *Observatory* magazine, which carries news of Royal Astronomical Society meetings and certain other matters, contains a Stop-Press item. Two telegrams had arrived. One was from Crommelin in Sobral: 'ECLIPSE SPLENDID'. The other, from

Eddington, was disappointing: 'THROUGH CLOUD. HOPEFUL'. The expeditions returned and began to analyse their data. The community waited.

Measurement and Error

The full details of both expeditions can be read in the account published in *Philosophical Transactions of the Royal Society* (1920). The main items of experimental equipment were two astrographic object glasses of about 10 inches in diameter, one from Oxford and one from the Royal Greenwich Observatory. Such lenses are specially designed to measure star positions over a relatively large piece of the sky, and were therefore ideal for the kind of experiment being done during the eclipse. The 'objectives' were removed from the observatories in which they were usually housed, and steel tubes were built to form temporary telescopes for the expeditions. Almost as an afterthought, it was decided to take a much smaller objective lens, 4 inches in diameter, to the Sobral site as a kind of backup.

The expeditions also took two large coelostats

(special mirrors used for solar observations). The reason for the mirrors was that no mechanical devices were available to drive the steel tubes containing the object glasses to compensate for the rotation of the Earth. The tubes had to be as long as the focal length of the lens – in this case, about 3.5 metres – so they were difficult to move, once set up. If a telescope is not moved by such a driver during the taking of a photograph, the stars move on the sky during the exposure and the images turn into streaks. In the eclipse experiment, the trick was to keep the telescope still, but to have it pointing downwards towards the coelostat, which reflected the light into the telescope lens. The mirror was much smaller than the tube (about 16 inches across), and a relatively simple clockwork device could be used to move it to correct for the Earth's rotation, instead of moving the whole telescope.

It is clear that both expeditions encountered numerous technical problems. The day of the eclipse arrived at Principe with heavy cloud and rain. Eddington was almost washed out, but near totality the Sun began to appear dimly through cloud and some photographic images could be taken. Most of these were unsuccessful, but the Principe mission did manage to return with two useable photographic plates. Sobral had better weather, but Crommelin had made a blunder during the setting up of his main telescope. He and his team had set the focus overnight before the eclipse, when there were plenty of bright stars around to check the optical performance of the telescope. However, when the day of the eclipse dawned and the temperature began to rise, his team watched with growing alarm as both the steel tube and the coelostat mirror began to expand with the heat. As a result, most of the main Sobral plates were badly blurred. On the other hand, the little 4-inch telescope taken as a backup performed very well, and the plates obtained with it were to prove the most convincing in the final analysis.

There were other problems too. The light deflection expected was quite small: less than two seconds of arc. But other things could cause a shifting of the stars' position on a photographic plate. For one thing, photographic plates can also expand and contract with changes in temperature. The emulsion used might not be particularly uniform. The eclipse plates might have been exposed under different conditions from the reference plates, and so on. The Sobral team in particular realised that, having risen during the morning, the temperature fell noticeably during totality, with the probable result that the photographic plates would shrink. The refractive properties of the atmosphere also change during an eclipse, leading to a false distortion of the images. And perhaps most critically of all, Eddington's expedition was hampered by bad luck even after the eclipse. Because of an imminent strike of the local steamship operators, his team was in danger of being completely stranded. He was therefore forced to leave early, before taking any reference plates of the same region of the sky with the same equipment. Instead he relied on one check plate made at Principe and others taken previously at Oxford. These were better than nothing, but made it impossible to check fully for systematic errors, and laid his results open to considerable criticism. All these problems had to be allowed for, and corrected if possible in the final stage of data analysis.

Scientific observations are always subject to errors and uncertainty of this kind. The level of this uncertainty in any experimental result is usually communicated in the technical literature by giving not just one number as the answer, but attaching to it another number called the 'standard error', an estimate of the range of possible errors that could influence the result. If the light deflection measured was, say, 1 arc second, then this measurement would be totally unreliable if the standard error were as large as the measurement itself, 1 arc second. Such a result would be presented as '1 ± 1' arc second, and nobody would believe it because the measured deflection could well be produced entirely by instrumental errors. In fact, as a rule of thumb, physicists never usually believe anything unless the measured number is larger than two standard errors. The expedition teams analysed their data, with Eddington playing the leading role, cross-checked with the reference plates, checked and double-checked their standard errors. Finally, they were ready.

Results and Reaction

A special joint meeting of the Royal Astronomical Society and the Royal Society of London was convened on 6 November 1919. Dyson presented the main results, and was followed by contributions from Crommelin and Eddington. The results from Sobral, with measurements of seven stars in good visibility, gave the deflection as 1.98 ± 0.16 arc seconds. Principe was less convincing. Only five stars were included, and the conditions there led to a much larger error. Nevertheless, the value obtained by Eddington was 1.61 ± 0.40. Both were within two standard errors of the Einstein value of 1.74 and more than two standard errors away from either zero or the Newtonian value of 0.87.

The reaction from scientists at this special meeting was ambivalent. Some questioned the reliability of statistical evidence from such a small number of stars. This scepticism seems in retrospect to be entirely justified. Although the results from Sobral were consistent with Einstein's prediction, Eddington had been careful to remove from the analysis all measurements taken with the main equipment – the astrographic telescope – and to use only the results from the 4-inch. As I have explained, there were good grounds for this, because of problems with the focus of the larger instrument. On the other hand, these plates yielded a value for the deflection of 0.93 seconds of arc, very close to the Newtonian prediction. Some suspected Eddington of cooking the books by leaving these measurements out. Others, such as Ludwick Silberstein, admonished the audience. Silberstein pointed a finger at the portrait of Newton that hangs in the meeting room, and warned: 'We owe it to that great man to proceed very carefully in modifying or retouching his Law of Gravitation.' On the other hand, the eminent Professor J.J. Thomson, discoverer of the electron and Chair of the meeting, was convinced: 'This is the most important result obtained in connection with the theory of gravitation since Newton's day.'

Einstein himself had no doubts. He had known about the results from the English expeditions before the formal announcement in November 1919. On 27 September, he had written an excited postcard to his mother:

. . . joyous news today. H.A. Lorentz telegraphed that the English expeditions have actually measured the deflection of starlight from the Sun.

He later down-played his excitement in a puckish remark about his friend and colleague, the physicist Max Planck:

He was one of the finest people I have ever known . . . but he didn't really understand physics, [because] *during the eclipse of 1919 he stayed up all night to see if it would confirm the bending of light by the gravitational field. If he had really understood* [the general theory of relativity], *he would have gone to bed the way I did.*

In 1922, another eclipse, viewed this time from Australia, yielded not a handful, but scores of measured position-shifts and much more convincing statistical data. But even so, the standard error on these later measurements was of similar size, around 0.20 arc seconds. Measurements of this kind using optical telescopes to measure light deflection continued until the 1950s, but never increased much in accuracy because of the fundamental problems in observing stars through the Earth's atmosphere. More recently, similar measurements have been made, not using optical light but radio waves. These have the advantage that they are not scattered by the atmosphere like optical light is. The light-bending measurement for radio sources rather than stars can be made almost at will, without having to wait for an eclipse. For example, every year the distant quasar 3C279 passes behind the Sun, producing a measurable deflection. These measurements confirm the Einstein prediction, and it is now accepted by the vast majority of physicists that light is bent in the manner suggested by the general theory of relativity.

Moreover, other predictions of the general theory also seem to fit with recent observations: gravitational time dilation; gravitational redshift; the orbital spin of the binary pulsar; the perihelion advance of Mercury; and so on. All these have been measured and are consistent with Einstein's theory. Nowadays, astronomers even use the bending as a measurement tool, so confident

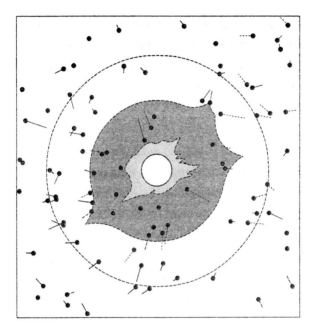

Figure 6. Changes in star positions recorded by Campbell and Trumper during the eclipse of 1922. The plates obtained in 1919 are of insufficient quality to reproduce well, but this example illustrates the kind of data obtained. The Sun during the eclipse is represented by the circle in the centre of the diagram, surrounded by a representation of the coronal light. Images too close to the corona cannot be used. The recorded displacements of other stars are represented by lines (not to scale).

are they of its theoretical basis. For example, the focusing effect of distant galaxies can form multiple images of background objects such as quasars, a phenomenon called gravitational lensing. This can be used to weigh galaxies and perhaps even tell us whether the Universe is finite or infinite.

The eclipse expeditions of 1919 certainly led to the eventual acceptance of Einstein's general theory of relativity in the scientific community. This theory is now an important part of the training of any physicist and is regarded as the best we have for describing the various phenomena attributable to the action of gravity. The events of 1919 also established Einstein, rightly, as one of the century's greatest intellects. But it was to do much more than that, propelling him from the rarefied world of theoretical physics into the domain of popular culture. How did this happen?

Einstein the Icon

Einstein, and his theory of relativity, had appeared in newspapers before, mainly in the German-speaking world. He had himself written an article for *Die Vossische Zeitung* in 1914. But he had never experienced anything like the press reaction to the announcements at the Royal Society meeting in 1919. Indeed, as Abraham Pais notes in his superb biography of Einstein, the *New York Times* index records no mention at all of Einstein until 9 November 1919. From then until his death in 1955, not a year passed without a mention of Einstein's name.

Some of the initial attention also rubbed off on Eddington. He ran a series of lectures in Cambridge on Einstein's theory. Hundreds turned up and the lectures were packed. Eddington became one of the foremost proponents of the new theory in England, and went on to inspire a generation of astrophysicists in Cambridge and beyond. But this was nothing compared to what happened to Albert Einstein.

The London *Times* of 7 November 1919 carried a long article about the Royal Society meeting, headlined 'REVOLUTION IN SCIENCE. NEW THEORY OF THE UNIVERSE'. Two days later, the *New York Times* appeared with the headline 'LIGHTS ALL ASKEW IN THE

HEAVENS'. But these splashes were not to be short-lived. Day after day, the global media ran editorials and further features about Einstein and his theory. The man himself was asked to write an article for the London *Times*, an offer he accepted 'with joy and gratefulness'. Gradually, the press reinforced the role of Einstein as genius and hero, taking pains to position him on one side of an enormous intellectual gulf separating him from the common man. He emerged as a saintly, almost mythical character who was accorded great respect by scientists and non-scientists alike.

As the years passed, his fame expanded further still, reaching into parts of popular culture that scientists had never occupied before. Einstein was invited to appear in Variety at the London Palladium (doing what, one can only guess). He featured in popular songs, films and advertisements. Eventually, this attention wore him down. Towards the end of his life he wrote to a friend:

Because of the peculiar popularity which I have acquired, anything I do is likely to develop into a ridiculous comedy. This means that I have to stay close to home and rarely leave Princeton.

No scientist working today would begrudge the fame that settled on Einstein. His achievements were stunning, with all the hallmarks of genius stamped upon them. But while his scientific contributions were clearly a necessary part of his canonisation, they are not sufficient to explain the unprecedented public reaction.

One of the other factors that played a role in this process is obvious when one looks at the other stories in the London *Times* of 7 November 1919. On the same page as the eclipse report, one finds the following headlines: 'ARMISTICE AND TREATY TERMS'; 'GERMANS SUMMONED TO PARIS'; 'RECONSTRUCTION PROGRESS'; and 'WAR CRIMES AGAINST SERBIA'. To a world wearied by a terrible war, and still suffering in its aftermath, this funny little man and his crazy theories must have been a welcome distraction, even if his ideas themselves went way over the heads of ordinary people. Here too was token of a much-needed reconciliation between Britain and Germany. In his *Times* article, Einstein stressed that science cuts across mere national boundaries, hinting that if politicians behaved more like scientists there would be no more pointless destruction on the scale that Europe had just experienced.

But there was a more human side to the Einstein phenomenon. The image of the man himself seemed to fit the public idea of what a scientist should be. He was a natural born cliché, the stereotypical absent-minded professor. With his kindly, instantly recognisable face, gentle personal manner and vaguely shambolic appearance, he looked like everyone's favourite uncle. Though a genius, he lacked arrogance. His political views, such as his widely publicised pacifism, meant there was always a distance between him and the establishment that had led Europe into disastrous conflict. Perhaps even the overthrow of Newton's theory of gravity was seen as a healthy kick up the backside of the old order. He filled a role that the public needed.

The only other physicist to have been afforded this kind of global megastardom is the British theoretician, Stephen Hawking. Here too, scientific achievements are only part of the story. Hawking's emergence as a cultural icon has many similarities, though the Hawking cliché is different: the brilliant brain trapped in a crippled body. (Hawking has suffered from Motor Neurone Disease since his early twenties.) One difference, however, is that, contrary to popular belief, Hawking is a somewhat peripheral figure in the world of modern physics. Posterity has not yet had time to confirm his place in the physicists' hall of fame to the same extent that it has for Einstein.

The Press, Science, and Truth

A few years ago, I was interviewed for a BBC programme about cosmology. Appropriately enough, the programme was called *Big Science*. The interviewer asked me a simple question: if there is a scientific controversy in a field so far removed from public understanding, how could ordinary people decide which side is right? I was stumped. Much to the amusement of my friends, the programme went out with a sequence of me scratching my head and saying I didn't know the answer. I have thought long and hard about this since then, but still don't have an answer. I have, however, decided that it is the wrong question.

Science does not deal with 'rights' and 'wrongs'. It deals instead with descriptions of reality that are either 'useful' or 'not useful'. Newton's theory of gravity was not shown to be 'wrong' by the eclipse expedition. It was merely shown that there were some phenomena it could not describe, and for which a more sophisticated theory was required. But Newton's theory still yields perfectly reliable predictions in many situations, including, for example, the timing of total solar eclipses. When a theory is shown to be useful in a wide range of situations, it becomes part of our standard model of the world. But this doesn't make it true, because we will never know whether future experiments may supersede it. It may well be the case that physical situations will be found where general relativity is supplanted by another theory of gravity. Indeed, physicists already know that Einstein's theory breaks down when matter is so dense that quantum effects become important. Einstein himself realised that this would probably happen to his theory.

Putting together the material for this book, I was struck by the many parallels between the events of 1919 and coverage of similar topics in the newspapers of 1999. One of the hot topics for the media in January 1999, for example, was the discovery by an international team of astronomers that distant exploding stars called supernovae are much fainter than had been predicted. To cut a long story short, this means that these objects are thought to be much further away than expected. The inference then is that not only is the Universe expanding, but it is doing so at a faster and faster rate as time passes. In other words, the Universe is accelerating. The only way that modern theories can account for this acceleration is to suggest that there is an additional source of energy pervading the very vacuum of space. These observations therefore hold profound implications for fundamental physics.

As always seems to be the case, the press present these observations as bald facts. As an astrophysicist, I know very well that they are far from unchallenged by the astronomical community. Lively debates about these results occur regularly at scientific meetings, and their status is far from established. In fact, only a year or two ago, precisely the same team was arguing for exactly the opposite conclusion based on their earlier data. But the media don't seem to like representing science the way it actually is, as an arena in which ideas are vigorously debated and each result is presented with caveats and careful analysis of possible error. They prefer instead to portray scientists as priests, laying down the law without equivocation. The more esoteric the theory, the further it is beyond the grasp of the non-specialist, the more exalted is the priest. It is not that the public want to know – they want not to *know* but to *believe*.

Things seem to have been the same in 1919. Although the results from Sobral and Principe had not then received independent confirmation from other experiments, in much the same way as the new supernova experiments have not, they were still presented to the public at large as being definitive proof of something very profound. That the eclipse measurements later received confirmation is not the point. This kind of reporting can elevate scientists, at least temporarily, to the priesthood, but does nothing to bridge the ever-widening gap between what scientists do and what the public think they do.

As we enter a new millennium, science continues to expand into areas still further beyond the comprehension of the general public. Particle physicists want to understand the structure of matter on tinier and tinier scales of length and time. Astronomers want to know how stars, galaxies and life itself came into being. But it is not only the *theoretical* ambition of science that is getting bigger: experimental tests of modern particle theories require methods capable of probing objects a tiny fraction of the size of the nucleus of an atom. With devices such as the Hubble Space Telescope, astronomers can gather light that comes from sources so distant that it has taken most of the age of the Universe to reach us. But extending these experimental methods still further will require yet more money to be spent. The further science reaches beyond the general public, the more it relies on their taxes.

Many modern scientists themselves play a dangerous game with the truth, pushing their results one-sidedly into the media as part of the cutthroat battle for a share of scarce research funding. There may be short-term rewards, in grants and TV appearances, but in the long run the impact on the relationship between science and society can only be bad. The public responded to Einstein with unqualified admiration, but Big Science later gave the world nuclear weapons. The distorted image of the scientist-as-priest is likely to lead only to alienation and further loss of public respect. Science is not a religion, and should not pretend to be one.

Further Reading

By far the best biography of Albert Einstein is:

Pais, A., *'Subtle is the Lord . . . ' The Science and the Life of Albert Einstein*, Oxford: Oxford University Press, 1992.

For a less technical account of his life and work, see:

Hey, T., and Walters, P., *Einstein's Mirror*, Cambridge: Cambridge University Press, 1997.

The life of Eddington is described in:

Douglas, A.V., *The Life of Arthur Stanley Eddington*, London: Thomas Nelson & Sons, 1957.

Chandrasekhar, S., *Eddington. The Most Distinguished Astrophysicist of His Time*, Cambridge: Cambridge University Press, 1983.

The scientific results of the Sobral–Principe expeditions, including a reproduction of one of the original plates, are published in:

Dyson, F.W., Eddington, A.S., and Davidson, C., *Philosophical Transactions of the Royal Society of London*, Series A, Vol. 220, pp. 291–333, London: 1920.

A detailed technical discussion of light deflection measurements and other tests of Einstein's theory can be found in:

Bertotti, B., Brill, D.R., and Krotkov, R., 'Experiments on Gravitation', in *Gravitation: an Introduction to Current Research*, ed. L. Witten, pp. 1–48, New York: John Wiley & Sons, 1962.

A more sceptical take on the Eddington expeditions can be found in:

Collins, H.M., and Pinch, T., *The Golem: What You Should Know About Science* (2nd edition), Cambridge: Cambridge University Press, 1998.

A variety of wonderful pictures, movies and general information about eclipses can be found on the Internet at:

http://sunearth.gsfc.nasa.gov/eclipse/eclipse.html

Key Ideas

Bending of Light

Light follows the shortest path between two points. In a flat space, this is a straight line, but in a curved space, such a path will also be curved. This leads to the possible measurement of light-bending by a massive body like the Sun. Far from the Sun, space is almost flat, so light travels in a near-perfect straight line. Near the Sun, however, space is curved, so a light ray will be deflected slightly off its straight-line path. That means that if you observe stars behind the Sun, their positions will be slightly different from their positions when the Sun is out of the picture. The effect is small, but measurable. Measurements of the bending of starlight provide strong evidence that space is indeed curved.

Curved Space

One of the consequences of Einstein's general theory of relativity is that massive bodies distort the space around them. In the Newtonian world, space is 'flat', meaning that the laws of geometry and trigonometry always apply. The angles of a triangle add up to 180 degrees, and so on. According to

Einstein, the space around a body is not flat, so these rules can be violated. As a simple analogy, think of a flat rubber sheet stretched out horizontally. This is flat Newtonian space, although with two dimensions rather than three. Now put a heavy weight on the sheet. The sheet will be distorted downwards near the weight, producing a curved space such as the general relativity predicts.

Einstein's Theory of Relativity

The theory of relativity is founded on the idea that only relative motion can be measured. The consequences of this notion are profound, and shatter the Newtonian conception of the world. Both space and time are no longer absolutes, but depend on the state of motion of whoever is measuring them. Moving clocks run slow; moving rulers appear shorter. Time and space are no longer absolute, but are fused together in a four-dimensional 'space-time'. These conclusions were established in the 'special' theory in 1905, which dealt only with motion at constant speed. Later work on the 'general' theory of relativity (1915) brought acceleration and gravity into the picture, leading to the idea that space-time can be distorted. In the general theory, motion in space-time affects the properties of space-time. Far from being a passive stage, as it is in Newtonian mechanics, the general theory of relativity depicts space-time as inherently dynamical.

Light and Energy

Light is a form of electromagnetic radiation, just as radio waves and X-rays are. Such radiation is made of small packets ('photons') of pure energy. Although these packets behave like particles in some ways, unlike material particles they have no mass. According to Newtonian mechanics, they therefore should not feel the effect of gravity. In relativity, however, there is a kind of equivalence between energy and mass expressed by the famous relation $E=mc^2$. In Einstein's general theory, this means that light does indeed feel the effects of gravitation, just as mass does.

Newtonian Mechanics

Sir Isaac Newton played a major role in the scientific revolutions of the 17th century. In the famous *Principia* he codified the mathematical laws of motion and also formulated the law of universal gravitation. The stage on which Newton's mechanics are acted out is an absolute one. 'Space' is an unchanging backdrop against which bodies move and interact, and which is described perfectly by Euclid's laws of geometry. Likewise, time is absolute, universal and independent of motion of bodies in space.

Path of Totality

During a solar eclipse, the region of the Earth's surface where the Sun's light is completely blotted out at any time is quite small – a few hundred miles across. As the Earth rotates, this dark spot moves, sweeping out a narrow region called the *path of totality*. On the path of totality one can see a total solar eclipse; away from it, either no eclipse will be seen, or it will only be partial.

Types of Eclipse

A *solar* eclipse occurs when the Moon passes across a direct line-of-sight from the Earth to the Sun, casting a shadow on the Earth. If the geometry of the Earth-Moon-Sun system is just right when this happens, and one observes from the right place on Earth, the eclipse can be *total*, i.e. the Sun's disk is completely obscured by that of the Moon. A more general situation is a *partial* eclipse, in which the Moon does not completely blot out the Sun's light. Sometimes, a partial eclipse can be *annular*, when the Moon's disk fits entirely inside the Sun's. A *lunar* eclipse occurs when the Earth instead passes between Sun and Moon, casting a shadow on the Moon and changing its colour.

Other titles available in the Postmodern Encounters series from Icon/Totem

Derrida and the End of History
Stuart Sim
ISBN 1 84046 094 6
UK £2.99   USA $7.95

What does it mean to proclaim 'the end of history', as
several thinkers have done in recent years? Francis
Fukuyama, the American political theorist, created a
considerable stir in *The End of History and the Last Man*
(1992) by claiming that the fall of communism and the
triumph of free market liberalism brought an 'end of
history' as we know it. Prominent among his critics has
been the French philosopher Jacques Derrida, whose
Specters of Marx (1993) deconstructed the concept of
'the end of history' as an ideological confidence trick, in
an effort to salvage the unfinished and ongoing project
of democracy.

Derrida and the End of History places Derrida's claim
within the context of a wider tradition of 'endist' thought.
Derrida's critique of endism is highlighted as one of his
most valuable contributions to the postmodern cultural
debate – as well as being the most accessible entry to
deconstruction, the controversial philosophical
movement founded by him.

Stuart Sim is Professor of English Studies at the
University of Sunderland. The author of several works on
critical and cultural theory, he edited *The Icon Critical
Dictionary of Postmodern Thought* (1998).

Foucault and Queer Theory
Tamsin Spargo
ISBN 1 84046 092 X
UK £2.99   USA $7.95

Michel Foucault is the most gossiped-about celebrity of French poststructuralist theory. The homophobic insult 'queer' is now proudly reclaimed by some who once called themselves lesbian or gay. What is the connection between the two?

This is a postmodern encounter between Foucault's theories of sexuality, power and discourse and the current key exponents of queer thinking who have adopted, revised and criticised Foucault. Our understanding of gender, identity, sexuality and cultural politics will be radically altered in this meeting of transgressive figures.

Foucault and Queer Theory excels as a brief introduction to Foucault's compelling ideas and the development of queer culture with its own outspoken views on heteronormativity, sado-masochism, performativity, transgender, the end of gender, liberation-versus-difference, late capitalism and the impact of AIDS on theories and practices.

Tamsin Spargo worked as an actor before taking up her current position as Senior Lecturer in Literary and Historical Studies at Liverpool John Moores University. She writes on religious writing, critical and cultural theory and desire.

Nietzsche and Postmodernism
Dave Robinson

ISBN 1 84046 093 8
UK £2.99   USA $7.95

Friedrich Nietzsche (1844–1900) has exerted a huge
influence on 20th century philosophy and literature – an
influence that looks set to continue into the 21st century.
Nietzsche questioned what it means for us to live in our
modern world. He was an 'anti-philosopher' who
expressed grave reservations about the reliability and
extent of human knowledge. His radical scepticism
disturbs our deepest-held beliefs and values. For these
reasons, Nietzsche casts a 'long shadow' on the complex
cultural and philosophical phenomenon we now call
'postmodernism'.

Nietzsche and Postmodernism explains the key ideas of
this 'Anti-Christ' philosopher. It then provides a clear
account of the central themes of postmodernist thought
exemplified by such thinkers as Derrida, Foucault,
Lyotard and Rorty, and concludes by asking if Nietzsche
can justifiably be called the first great postmodernist.

Dave Robinson has taught philosophy for many years. He
is the author of Icon/Totem's introductory guides to
Philosophy, Ethics and Descartes. He thinks that
Nietzsche is a postmodernist, but he's not sure.

Baudrillard and the Millennium
Christopher Horrocks

ISBN 1 84046 091 1
UK £2.99   USA $7.95

'In a sense, we do not believe in the Year 2000', says French thinker Jean Baudrillard. Still more disturbing is his claim that the millennium might not take place. Baudrillard's analysis of 'Y2K' reveals a repentant culture intent on storing, mourning and laundering its past, and a world from which even the possibility of the 'end of history' has vanished. Yet behind this bleak vision of integrated reality, Baudrillard identifies enigmatic possibilities and perhaps a final ironic twist.

Baudrillard and the Millennium confronts the strategies of this major cultural analyst's encounter with the greatest non-event of the postmodern age, and accounts for the critical censure of Baudrillard's enterprise. Key topics, such as natural catastrophes, the body, 'victim culture', identity and Internet viruses, are discussed in reference to the development of Jean Baudrillard's millenarian thought from the 1980s to the threshold of the Year 2000 – from simulation to disappearance.

Christopher Horrocks is Senior Lecturer in Art History at Kingston University in Surrey. His publications include *Introducing Baudrillard* and *Introducing Foucault*, both published by Icon/Totem. He lives in Tulse Hill, in the south of London.